2023 Quantum Computing Applications in Chemical and Biochemical Engineering Workshop

Lyngby, Denmark
10-12 May 2023

ISBN: 978-1-7138-8675-4

Printed from e-media with permission by:

Curran Associates, Inc.
57 Morehouse Lane
Red Hook, NY 12571

Printed with permission by Curran Associates, Inc. (2024)

For permission requests, please contact AIChE
at the address below.

AIChE
120 Wall Street, FL 23
New York, NY 10005-4020

Phone: (800) 242-4363
Fax: (203) 775-5177

www.aiche.org

Additional copies of this publication are available from:

Curran Associates, Inc.
57 Morehouse Lane
Red Hook, NY 12571 USA
Phone: 845-758-0400
Fax: 845-758-2633
Email: curran@proceedings.com
Web: www.proceedings.com

TABLE OF CONTENTS

KEYNOTE 4: FRED CHONG, UNIVERSITY OF CHICAGO

IBM QISKIT WORKSHOP: INCLUDES INTRODUCTORY AND NATURE TUTORIALS

LIGHTNING PRESENATIONS & POSTER SESSION RECEPTION

INDUSTRY APPLICATIONS OF QUANTUM COMPUTING

Panel Discussion

Conference

Quantum Computing Applications Chemical and Biochemical Engineering Workshop

Wednesday, May 10, 2023 - 10:50am to 11:30am

The Novo Nordisk Foundation Quantum Computing Programme launched earlier this year with a mission to enable the development of fault tolerant quantum computing hardware to solve a problem of significance in the life sciences. In this talk I will outline the strategy we are developing for the Applications & Algorithms area. In particular, I will focus on our search for use cases that constitute problems of significance and at the same might be amenable to solution with quantum algorithms.

Prospects of Quantum Computing in Chemistry, Molecular Biology, and Materials Science

Conference

Quantum Computing Applications Chemical and Biochemical Engineering Workshop

Wednesday, May 10, 2023 - 10:30am to 10:50am

Authors

Reiher, M. - Presenter, ETH Zurich

Many problems in molecular science and condensed-phase systems, which are both governed by the dynamics of electrons and atomic nuclei, demand an explicit quantum mechanical description. In such quantum problems, the representation of wave functions grows exponentially with system size, which poses a severe restriction on traditional approaches. However, such quantum problems should naturally benefit from digital quantum simulation on a number of logical qubits, as this would scale only linearly with system size. In recent years, we have considered quantum computing applications in molecular biology, catalysis, and physical chemistry in general, with a focus on how and where to establish a quantum advantage in these areas (see list of references below). In my talk, I will elaborate on the potential benefits of quantum computing in these application areas, especially when compared to state-of-the-art traditional approaches.

The Novo Nordisk Foundation Quantum Computing Programme: Visions for Algorithms & Applications

Conference

Quantum Computing Applications Chemical and Biochemical Engineering Workshop

Wednesday, May 10, 2023 - 10:10am to 10:30am

Authors

Solomon, G. - Presenter, Niels Bohr Institute

The Novo Nordisk Foundation Quantum Computing Programme launched earlier this year with a mission to enable the development of fault tolerant quantum computing hardware to solve a problem of significance in the life sciences. In this talk I will outline the strategy we are developing for the Applications & Algorithms area. In particular, I will focus on our search for use cases that constitute problems of significance and at the same might be amenable to solution with quantum algorithms.

Hybrid Quantum Chemistry on Hybrid Quantum Computers - (HQC)2

Conference

Quantum Computing Applications Chemical and Biochemical Engineering Workshop

Wednesday, May 10, 2023 - 9:50am to 10:10am

Authors

Sauer, S. P. A. - Presenter, University of Copenhagen, Copenhagen

Coriani, S., Technical University of Denmark

Kongsted, J., University of Southern Denmark

The process of rational design of tailor-made molecules possessing desired and specific properties may be brought to revolution, if the relevant molecular properties can be accurately predicted on a reasonable time scale. Even if the necessary quantum chemical methods to achieve such molecular property calculations are well known to quantum chemists, carrying out these calculations on classical computers for large biological molecules in their natural environments is today too slow, if not impossible. This scenario could, however, completely change when large-scale quantum computers become available. They are expected to provide the opportunity to more accurately calculate the outcomes of complex chemical and biochemical processes. Before this realization, however, significant challenges need to be overcome since current quantum chemical software cannot simply be ported to quantum computers. Thereby, specific methodological and algorithmic developments tailored for quantum computers are thus in high need in order to facilitate the transition of molecular property calculations from classical to quantum computers.

This is exactly the goal of our new research project supported by the Novo Nordisk Foundation. In this presentation, we will discuss the specific types of quantum chemical methods and calculations we intend to implement on quantum computers and which quantum algorithms we will employ for this goal.

Panel Discussion

Wednesday, May 10, 2023 - 1:50pm to 2:30pm

The Novo Nordisk Foundation Quantum Computing Programme launched earlier this year with a mission to enable the development of fault tolerant quantum computing hardware to solve a problem of significance in the life sciences. In this talk I will outline the strategy we are developing for the Applications & Algorithms area. In particular, I will focus on our search for use cases that constitute problems of significance and at the same might be amenable to solution with quantum algorithms.

Quantum simulation of battery materials using ionic pseudopotentials

Wednesday, May 10, 2023 - 1:30pm to 1:50pm

Authors

Arrazola, J. M. - Presenter, Xanadu

We introduce a quantum algorithm that uses pseudopotentials to reduce the cost of simulating periodic materials on a quantum computer. We address the challenge of incorporating the complexity of pseudopotentials into quantum simulations by developing highly-optimized compilation strategies for the qubitization of the Hamiltonian. This includes a linear combination of unitaries decomposition that leverages the form of separable pseudopotentials. We estimate the computational cost of applying our algorithm to simulating three lithium-excess cathode materials for batteries: lithium manganese oxide, lithium nickel-manganese oxide, and lithium manganese oxyfluoride. Our optimized compilation strategies result in a pseudopotential-based quantum algorithm with a total runtime four orders of magnitude lower than the previous state of the art for a fixed target accuracy.

Discrete nonlinear optimization: Modeling and solutions via novel hardware and decomposition algorithms

Wednesday, May 10, 2023 - 1:10pm to 1:30pm

Authors

Bernal Neira, D. E. - Presenter, Universities Space Research Association – Research Institute of Advanced Computer Science

Optimization problems arise in different areas of Logistics, Manufacturing, Process Systems Engineering (PSE), and Energy Systems Engineering, and solving these problems efficiently is essential for addressing important industrial applications.

Quantum computers have the potential to efficiently solve challenging nonlinear and combinatorial problems. However, available quantum computers cannot efficiently address practical problems; they are limited to small sizes and do not handle constraints well. In this talk and tutorial, we present the modeling strategy of problems as Mixed-Integer Nonlinear Programs (MINLP), explain some of the approaches that quantum computers use to solve Quadratic Unconstrained Binary Optimization (QUBO) problems, and propose hybrid classical-quantum algorithms to solve a class of MINLP, mixed-binary quadratically constrained programs (MIQCP) and apply decomposition strategies to break them down into QUBO subproblems that can be solved by quantum computers. This approach is based on a copositve optimization reformulation of the optimization problems, integrated within a cutting plane algorithm. The overall algorithm provides optimality convergence guarantees, yet it is robust to suboptimal solutions of the QUBO problems, which are usually provided by the hardware-based approaches (e.g., Quantum Annealing) to QUBO (arXiv:2207.13630).

We will also cover different approaches to formulating and solving Quadratic Unconstrained Binary Optimization (QUBO) problems through unconventional computation methods, including but not limited to Quantum algorithms, and discuss how these approaches lead to algorithms able to outperform classical solution approaches.

The Quantum for Life Center

Wednesday, May 10, 2023 - 12:50pm to 1:10pm

Authors

Christandl, M. - Presenter, University of Copenhagen

The Quantum for Life Center, funded by the Novo Nordisk Foundation, was kicked-off in 2022 with the mission investigate the impact of quantum computing to the life sciences. One year on, the team is still growing as well as solidifying collaborations within and with the outside world. I will give an overview over our activities, from theory to experiment, as well as give a deep-dive into a theory challenge that keeps us awake at night.

The journey from research to QC applications at Zapata

Wednesday, May 10, 2023 - 3:30pm to 4:00pm

Authors

Fontalvo, J. R. - Presenter, Zapata Computing

With a team of over 30 PhDs, quantum computing research is a fundamental activity at Zapata. In this presentation, we will describe some of our most recent technological advancements and how we are turning them into applications for industry use cases.â€œ

Presenter information: Jhonathan Romero was born and raised in Colombia, where he earned his B.Sc. and M.Sc. degrees in Chemistry. Early on, Jhonathan specialized in theoretical and computational Chemistry, which led him to join the Aspuru-Guzik group at Harvard University in 2014. Here his research in Chemical Physics was centered on developing variational quantum algorithms for quantum simulation and machine learning. He received his Ph.D. from Harvard University in 2019. As Director of Professional Services and Co-Founder, he leads Zapataâ€™s efforts to execute customer projects. He also participates in many science outreach initiatives, including Science Clubs Colombia.

Quantum algorithms for the dynamics of molecular and biological systems

Authors

Head-Marsden, K. - Presenter

Lossy, or open, quantum dynamics are ubiquitous in nature, but challenging to model due to the complexity of environmental interactions. While the idea of using quantum computing platforms to simulate these dynamical processes in either a digital or analog fashion has been around for decades, it wasnâ€™t until recently that the hardware has become sufficiently accessible to verify these methods. Here, I will give an overview of recent developments in the field of quantum computation and algorithms for modelling the dynamics of open quantum systems, with an emphasis on digital simulation, or gate-based, techniques applied to molecular and biological systems.

Panel Discussion

Thursday, May 11, 2023 - 11:20am to 11:40am

In this presentation, Tennin will cover QunaSys' recent research of fundamentally improving the VQE algorithm and developing of partial error correction algorithm for chemistry applications. QunaSys is expanding its community to Europe this spring. Tennin will share how QunaSys leverages their research and technology to assist companies in becoming quantum-ready

IQM - We build Quantum Computers (in Europe)

Thursday, May 11, 2023 - 11:00am to 11:20am

Authors

Eder, P. - Presenter, IQM Quantum Computers

IQM is the leading European vendor of quantum computers based on superconducting technology. IQM strives to build world-leading quantum computers for the well-being of humankind, now and for the future and executes its strategy in three phases, called foundation, advantage and disruption. A key step in the first phase is to quantum-accelerate Europeâ€™s HPC centers that currently face a range of challenges as a result of fierce world-wide competition in high-performance computation. IQM partners with European HPCs on their quest to innovate and stay competitive. IQMâ€™s portfolio includes on-premises quantum computers for best access and control of quantum effects and algorithms and application-specific Co-Design that targets an acceleration of quantum advantage, as well as education and training for all audiences.

Fermionic Adaptive Sampling Theory for Variational Quantum Eigensolvers

Thursday, May 11, 2023 - 10:40am to 11:00am

Authors

Majland, M. - Presenter, Kvantify

Zinner, N. T., Kvantify and Aarhus University

ADAPT-VQE suffers from a significant measurement overhead when estimating the importance of operators in the wave function. In this talk, we will discuss Fermionic Adaptive Sampling Theory VQE (FAST-VQE), a method for selecting operators based on importance metrics solely derived from the populations of Slater determinants in the wave function. Our method mitigates measurement overheads for ADAPT-VQE as it is only dependent on the populations of Slater determinants which can simply be determined by measurements in the computational basis.

Beyond VQE: advancing quantum computing applicability in chemistry

Thursday, May 11, 2023 - 10:20am to 10:40am

Authors

Yan, T. - Presenter, Quansys

In this presentation, Tennin will cover QunaSys' recent research of fundamentally improving the VQE algorithm and developing of partial error correction algorithm for chemistry applications. QunaSys is expanding its community to Europe this spring. Tennin will share how QunaSys leverages their research and technology to assist companies in becoming quantum-ready.

Paving the way for useful computational chemistry with the orbital-optimised variational quantum eigensolver

Thursday, May 11, 2023 - 10:00am to 10:20am

Authors

Knecht, S. - Presenter, Algorithmiq Ltd

In the past years, the capability of near-term quantum computers to address the electronic structure problem for chemical systems has been successfully demonstrated for toy-model systems such as, for example, N_2, H_2O and hydrogen chains H*n* (*n* â‰¤ 12).

Hence, surpassing the limited availability, reliability and capacity of current quantum hardware still poses a major challenge to date and calls for the development of efficiently scaling (embedded) quantum algorithms to treat systems of real chemical relevance, such as for the purpose of optimising binding affinities in the lead-optimisation step of the small- molecule drug discovery pipeline.

To this end, we present in this contribution the Variational Quantum Eigensolver (VQE) combined with the self-consistent field (SCF) approach to orbital optimization [1]. We will discuss both a classical [2] and a quantum embedding scheme [3] based on the polarizable embedding model and density-matrix embedding theory to account for subtle effects of the environment on the electronic structure of the embedded quantum mechanical system in its ground and electronically excited states. Moreover, binding affinities are often governed by molecular interactions which necessitate to consider not only static but also dynamical electron-electron correlations. We address these challenges by making use of two distinct frameworks for VQE-SCF that can be classified as *diagonalise-then-perturb* [4] and *perturb-and-diagonalise* [5], respectively.

[1] A. Fitzpatrick, A. Nykänen, N. W. Talarico, A. Lunghi, S. Maniscalco, G. García-Pérez, and S. Knecht, arXiv:2212.11405, 2022.

[2] E. Kjellgren, W. Talarico, M. Kadek, P. Reinholdt, J. Kongsted, S. P. A. Sauer, S. Coriani and S. Knecht, manuscript in preparation, 2023.

[3] M. Kadek, R. di Remigio Eikås, G. García-Pérez, and S. Knecht, work in progress,

2023.

[4] W. Talarico, A. Fitzpatrick, R. di Remigio Eikås, G. García-Pérez, and S. Knecht, manuscript in preparation, 2023.

[5] E. Kjellgren, J. Kongsted, S. P. A. Sauer, S. Coriani and S. Knecht, work in progress,

Greater Efficiency for Quantum Applications by Breaking Abstractions

Thursday, May 11, 2023 - 2:30pm to 3:00pm

Authors

Chong, F. - Presenter, University of Chicago

Classical computation has a significant role to play towards practical quantum computation. First, substantially greater circuit fidelities can be achieved through physics-aware compilation of circuits. Second, classical simulation of quantum circuits, especially special classes such as Clifford and Near-Clifford circuits, can substantially aid circuit optimization and initialization of quantum algorithms. Lastly, techniques such as circuit cutting can take advantage of a hybrid classical-quantum architecture to achieve computations beyond the scale of near-term quantum machines.

IBM Qiskit Workshop: includes introductory and nature tutorials

Thursday, May 11, 2023 - 3:15pm to 4:45pm

Authors

Tacchino, F. - Presenter, IBM

Baeumer, E. - Presenter, IBM Quantum

At the origins of quantum computation lies the challenge of simulating natural phenomena at their fundamental level. As the technology progresses, useful applications in fields such as chemistry, materials and life sciences become closer to reality.

In this tutorial IBM will first introduce you to basic quantum circuits and demonstrate how to implement them on actual quantum computers using Qiskit. In the second part IBM will then introduce and explore the functionalities of Qiskit Nature, a modular and open-source platform built to address a broad range of computational problems in the natural sciences with quantum algorithms and cloud-based quantum processors.

Quantum Computing Applications in Day-Ahead Market Optimisation

Thursday, May 11, 2023 - 5:35pm to 6:00pm

Authors

Brown, S. F. - Presenter, University of Sheffield

Wyatt-Jones, J. A., The University of Sheffield

Ejeh, J. O., The University of Sheffield

Quantum computing (QC) has attracted a great deal of research over recent years. This has been due to advances in hardware, software, solution algorithms, as well as its potential to show exponential speed-ups compared to equivalent classical computing. Although QC is still at its early stage, three key areas in chemical engineering have been identified to provide an initial advantage â€" computational chemistry, machine learning and optimisation (Ajageka and You, 2019).

An increasing number of chemical engineering problems can be posed as optimisation problems. One such problem relates to the design and scheduling of electrical energy storage (ESS) devices within power systems. ESS technologies have been shown to improve power system reliability, flexibility and security (Arghandeh et al., 2014; Ma et al., 2018), especially with the increasing reliance on intermittent renewable energy sources owing to their greenhouse gas emission reduction potential. This applies both in behind-the-meter and front-of-meter applications. The design and scheduling problem for the ESS device is important owing to high costs combined with the range of revenue streams â€" energy arbitrage, ancillary services, etc â€" such devices can have access to.

A range of existing optimisation models have been developed, suited to classical computers, which determine the optimal battery actions given a set of price signals for each revenue stream. However, current optimisation models are unsuitable for use on quantum processors. Hence, in this work we propose a reformulation of a mixed integer linear programming (MILP) model (Ejeh et al., 2022) for day-ahead market optimisation problem in a UK market, to a quantum unconstrained binary optimisation (QUBO) model. Using the Qiskit toolkit (Qiskit contributors, 2023*)*, the reformulated QUBO is solved using the quantum approximate optimisation algorithm (QAOA) with the minimum eigen optimiser. The results obtained from the MILP and QUBO models are analysed highlighting the performance, challenges and potential improvements for QC over current classical computing-based optimization approaches.

A Simple Visual Overview of Qiskit-Natures Components for Ab-Initio Calculations

Thursday, May 11, 2023 - 5:10pm to 5:35pm

Authors

Larrucea, J. - Presenter

Ab-initio calculations on atomic structures is per se a highly complex topic, which requires deep knowledge of mathematics and physics.
Luckily, over the decades, software and methods have been developed to allow for user friendlier workflows.
These calculations have been traditionally developed for classical von Neumann computing architectures.
The disruption of quantum computing technologies and their rapid development are now allowing us to perform the calculations of quantum mechanical systems natively on quantum hardware.
However, the added complexity of using quantum algorithms, does not help in understanding the new workflow, which involves a steep learning curve, even for experienced scientists with a traditional molecular modelling background.
In the present work, we present a visual overview of the workflow for ab-initio calculations using qiskit-nature, which is aimed at speeding up the learning process of scientists getting started on the field.

The Bonsai Algorithm: Mitigate Hardware Limitations By Growing Your Own Fermion-to-Qubit Mapping

Thursday, May 11, 2023 - 4:45pm to 5:10pm

Authors

Miller, A. - Presenter

In recent years, quantum computers have shown promising potential for solving the electronic structure problem in chemical systems. However, the limited connectivity of current quantum hardware still poses a major challenge for tackling complex chemical systems of actual relevance, as circuit overheads incurred by these constraints can quickly become unmanageable.

To address this challenge, we present the tree-based Bonsai algorithm [1], a formalism that takes into account connectivity constraints from the outset by â€?growingâ€? a fermion-to-qubit mapping tailored to architecture, reducing SWAP overhead. Applied to the heavy-hexagon topology used in IBM quantum computers, it produces mappings with a favourable Pauli weight scaling while avoiding the need for SWAP gates for single excitation operations. Furthermore, we relate the distribution of mode occupancy over qubits to the graph structure of the underlying mapping tree and formalize it by introducing the concept of the â€?delocalizationâ€? of a fermion-to-qubit mapping. With this understanding, we can directly control how we associate mode occupancy over qubits. Moreover, we show that paradigmatic mappings, such as the Jordan-Wigner, Parity, Brayvi-Kiteav, and ternary tree, are easily understood within this formalism.

[1] Miller, A., Zimborás, Z., Knecht, S., Maniscalco, S. & García-Peréz, G. The Bonsai algorithm: grow your own fermion-to-
qubit mapping. (2022)

Quantum Simulations beyond Electronic-Structure Methods: from Tensor Networks to Quantum Computing

Friday, May 12, 2023 - 10:40am to 11:00am

Authors

Baiardi, A. - Presenter, Laboratorium für Physikalische Chemie, ETH Zürich

Reiher, M., ETH Zurich

The simulation of quantum many-body systems is emerging as one of the most promising targets for quantum computing.1 In fact, quantum hardware promises to exponentially reduce the com- putational cost of these simulations compared to classical computers. In chemistry, quantum- computing algorithms have been mostly designed for time-independent electronic-structure cal- culations. However, other classes of many-body problems are relevant for chemical simulations. This includes vibrational-structure,2â€"4 coupled vibrational-electronic, 5,6 and time-dependent cal- culations.7 In this contribution, we will first describe methods that we developed, based on the density matrix renormalization group (DMRG) theory, for studying quantum-chemical many- body systems beyond electronic-structure problems. We will show how these methods can target systems that are hard challenges for alternative state-of-the-art quantum-chemical many-body methods. They set, therefore, the bar that quantum-computing algorithms must overcome to yield a practical quantum advantage in quantum-chemical simulations, as we will critically dis- cuss for the case of molecular vibrational-structure calculations.

Developing industry applications at Zapata

Friday, May 12, 2023 - 10:20am to 10:40am

Authors

Benfenati, F. - Presenter, University of l'Aquila

Zapata has been working with multiple customers on projects related to chemical engineering. In this presentation, we will describe some of those projects and the results we have achieved.â€œ

Presenter information: Francesco received his PhD in mechanical and industrial engineering at the University of Brescia, having spent 6 months at MIT as a visiting student, and his bachelors and masters in physics at the University of Milan. After focusing on molecular dynamics simulations, he started pursuing a career in quantum computing as a postdoctoral fellow at the University of Lâ€™Aquila, where he was collaborating with the IBM research team in Zurich, and later as a quantum software developer at Multiverse Computing. In his current role as quantum solutions engineer at Zapata Computing, he works with enterprises and universities to identify complex computational problems and find better ways to solve them using quantum computing and AI.

Quantum Simulation and Applications of Fault Tolerant QC

Friday, May 12, 2023 - 10:00am to 10:20am

Authors

Rubin, N. - Presenter, Google Quantum AI

An accurate assessment of how quantum computers can be used for chemical or materials simulation, especially their computational advantages, provides important context on how to deploy these future devices. To make this assessment we estimate the quantum resources required to sample the ground state of electronic structure Hamiltonians. This involves comparing various block encoding strategies, improving the block encoding constructions through symmetry-adapting, and ultimately compiling into the surface code to estimate physical qubit requirements and runtimes. To define a quantum simulation advantage we compare quantum runtime estimates to flagship classical simulations for a variety of challenging chemical and materials simulation problems.

Developing quantum solutions for pharmaceutical research and development

Friday, May 12, 2023 - 10:00am to 10:20am

Authors

Degroote, M. - Presenter, Boehringer Ingelheim

Quantum computers promise to impact industrial applications, for which quantum chemical calculations are required, by virtue of their high accuracy. This talk will explore the challenges and opportunities of applying quantum computers to drug design. I will discuss where they could transform industrial research and elaborate on what is needed to reach this goal.

Fault tolerant simulation of materials in 2nd quantization

Friday, May 12, 2023 - 9:40am to 10:00am

Authors

Rubin, N. - Presenter, Google Quantum AI

An accurate assessment of how quantum computers can be used for chemical or materials simulation, especially their computational advantages, provides important context on how to deploy these future devices. To make this assessment we estimate the quantum resources required to sample the ground state of electronic structure Hamiltonians. This involves comparing various block encoding strategies, improving the block encoding constructions through symmetry-adapting, and ultimately compiling into the surface code to estimate physical qubit requirements and runtimes. To define a quantum simulation advantage we compare quantum runtime estimates to flagship classical simulations for a variety of challenging chemical and materials simulation problems.

Quantum computing: technology and applications in the natural sciences

Friday, May 12, 2023 - 9:20am to 9:40am

Authors

Tacchino, F. - Presenter, IBM

Tavernelli, I., IBM

Over the last few decades, quantum information processing has emerged as a gateway towards new, powerful approaches to scientific computing. Quantum technologies are nowadays experiencing a rapid development and could lead to effective solutions in different domains including physics, chemistry, and life sciences, as well as optimisation, artificial intelligence, and finance. To achieve this goal, noise-resilient quantum algorithms together with error mitigation schemes have been proposed and implemented in hybrid workflows with the aim of improving the synergies between quantum and classical computational platforms. In this talk, I will review the state-of-the-art and recent progress in the field, both in terms of hardware and software, and present a broad spectrum of potential applications developed at IBM Research, with a focus on natural sciences and machine learning.

Photonic Quantum Computing - a status from the machine room

Friday, May 12, 2023 - 9:00am to 9:20am

Authors

Hoff, U. - Presenter, Technical University of Denmark

The superconducting qubit platforms have been leading the development for a while, but the quest for scalable error-corrected quantum computing remains an open one. Being an underdog for a long time, photonic measurement-based quantum computing based on highly entangled cluster states is now gaining momentum due to its favorable properties in terms of scalability.

In this talk, I will present the concept of measurement-based quantum computing on cluster states and provide a status on ongoing work at DTU on developing both near-term as well as full fault-tolerant photonic quantum computing systems in the context of national and European innovation and research projects.

AUTHOR INDEX